Independent Digital Intelligence

The Pinnacle of Human Invention

Bill Thurston

Independent Digital Intelligence

For information about the contents of this book, or permission for reproducing portions of this book, please email the author at thurstongroup@cox.net

Published in the United States

ISBN: 978-0-9886544-0-2

Table of Contents

Acknowledgements

Thanks to my wife, Debbie, who supported me and the amount of time I needed to create this book. Both she and I are glad the book is now written and published. We also want to thank you for purchasing my book.

Introduction

The next form of digital intelligence, after artificial intelligence (AI), is almost upon us. As you read this book, it will become increasing clear how extremely different this new form of digital intelligence is from what we use today. I call this new technology "IDI" which are initials for Independent Digital Intelligence. The operating system that uses this technology will be called the Independent Digital Intelligence Operating System or IDIOS for short. Any entity which uses the IDI technology will be called "IDB" which are initials for Independent Digital Body.

I do not want this book to be considered a textbook for engineers. A strong consideration in authoring this book is for anyone who is interested in the subject of the book to be able to understand the contents. I will use many real-life examples to explain new concepts. I will also have an extensive Table of Contents which will make it easy to find any of the topics and subtopics.

The reader with more technical knowledge can easily insert their knowledge into the subjects presented in the book.

My first book which I published in 2012 (titled "Really? Wow!") also addressed this new technology. It presented how this new technology will affect us in the next 25 and 50 years. In short, my first book addressed the "what" and "when" parts of this new technology whereas this book will address the "how" and "why" parts of this new technology. If you find interest in this book, you will also find interest in my first book.

The first time you read this book, it is important to read from the first chapter to the last chapter. The reason is that each chapter builds on the information presented in the previous chapter.

I hope you enjoy this book and if you would like to ask a question or communicate with the author or his team, use the email address on the back cover of the book.

I want to thank all the powerful people I learned from during my professional career. They formed a part of me. Given that, this book is totally my perceptions. Therefore, I find it easier to author this book in the first person. I hope that also makes it easier to read and enjoy.

It will be an exciting new world with our new tools and friends. They will make our world a better place.

The Technical Discussion

A few comments on the history of AI

Before jumping right into what I call the next generation of artificial intelligence, I want to make a few comments on the history of AI and how we perceive it today in 2021.

The two words "artificial" and "intelligence" when used together do not describe a particular thing. Just like the words "a car" or "an animal" or "a chair" do not describe a particular thing. To be totally clear, AI is simply a human-made program to emulate human-like intelligence (not dog or cat intelligence) and the associated abilities derived from that intelligence.

Artificial intelligence means something different depending on the timeframe of its development.

In the early 1900s there were actors on stage performing as artificial humans. A good example most of us know about is the Tinman and the Scarecrow in the "Wizard of Oz". Although the Tinman and Scarecrow acted like humans, people were not using the word "artificial intelligence" at that time. But that was the reality of AI in the early 1900s.

In the early 1960s, mainframe computers were showing up everywhere including many universities. When

I was attending the University of California in Berkeley, my computer experience went something like this:

- Go to the computer lab and get a coding form tablet.
- Go home and create your program in a language like Fortran.
- Go back to the computer lab and put your coding form into the "in" basket for the keypunch operators. Go back home.
- Next day, go to the computer lab and hope your key punched stack of cards were in the keypunch operators "out" basket.
- If your stack was there, you could now take them to another "in" basket for the special computer operators who would compile and run your program. Go back home.
- Next day, go back to the computer lab and hopefully in the "out" basket would be your cards and a large piece of paper from a "chain printer" with the results from your program.

Because in the 1960s very few people had seen a computer or written a computer program, they considered these computers to be "smart" like people and therefore considered people working with computers were working with an artificial form of intelligence. That was the reality of AI in 1964.

Today, in 2021, there are some very serious forms of AI that are available in everyday life. These products have humanlike features:

- Home automation
- Food Ordering
- Songs and Music
- Latest Updates about sports
- Messages and Hands-Free Calling
- Enterprise and Business Solutions
- Different kinds of virtual personal assistants
- Calendar Connection
- Creating Lists
- Alarms and Reminders
- News and Weather
- Q and A
- Recipes
- Etc.

Probably the best new example is the AI in our cars and trucks. They can parallel park better than most humans and back up with a trailer better than most humans. People who use the hands-free mode of driving place a lot of confidence in those programmers who create those hands-free programs.

My prediction is this AI activity will continue to develop with more sophisticated computer programs being written until IDI is developed and brought to market in its many forms.

The words to describe IDI, IDIOS and IDB

What are the right words to describe something that does not exist? That is a hard one!

The actual name of something is not important but your mental picture associated with the name is especially important.

That is why I chose the words "Independent Digital Intelligence", "Independent Digital Intelligence Operating System", and "Independent Digital Body" as names for these new entities. These words make these entities become partly clear just because we know the general meaning of each of the words. Had I just made up a new word and tried to describe the meaning of that word, it would be much more difficult to understand.

There are only five terms in the definition of these entities. They are the terms "digital", "intelligence", "independent", "body", and "operating system". It is important to take the time to clarify the meaning of each of them.

Defining the term "digital"

Digital means representing information in discrete states such as "on" and "off" or "1" and "0". Most light switches have two states or positions which are for having the light on or off. All digital computers today are called binary digital computers because they use many small switches that have only two states or positions.

Here is a simple example of the number of switches needed to create 16 GBs (gigabytes) of memory that you could find in your computer. It would take about 128,000,000,000 switches.

These switches are exceedingly small and are contained in things called miniature electronic devices. For these switches to work properly, they must be connected to an electricity source. Mobile devices get their electrical power from batteries or solar cells. Most personal computers and larger commercial computers get their electrical power from our universal power grid by being plugged into a wall plug (115 VAC or 220 VAC).

All IDBs require electricity.

Defining the term "intelligence"

Intelligence has been defined in many ways: the capacity for logic, understanding, self-awareness, learning, emotional knowledge, reasoning, planning, creativity, critical thinking, and problem-solving. More generally, it can be described as the ability to perceive or infer information, and to retain it as knowledge to be applied towards adaptive behaviors within an environment or context.

It is not merely book learning, a narrow academic skill, or test-taking smarts. Rather, it reflects a broader and deeper capability for comprehending our surroundings. Doing things like "catching on," "making sense" of things, or "figuring out" what to do.

For the purposes of having a mutual understanding during the discussion in this book, I will use this definition:

"Intelligence is the ability to acquire and apply knowledge and skills."

Although this is the most complicated part of the IDI technology, it is quite easy to explain. It is simply what is contained in the combination of the CPU (Central Processing Units) and computer operating system (OS). The equivalent part of the human would be the human brain.

In the existing digital world, some of the top producers of CPUs (Central Processing Units) are Intel, AMD, NVIDIA, and Qualcomm.

Some top computer operating systems are Android, Windows, IOS, MAC OS, and Linux.

At least in the beginning, IDBs will use existing CPUs. The IDIOS will be a unique operating system and IDBs will only use this operating system. In the future, the IDI technology will have its own unique CPU tailored for the IDI technology.

Just like existing operating systems, the IDI technology will be written in a computer language like “C” or “C++”.

Defining the term "independent"

Independent is the key difference between existing operating systems and the IDIOS.

Independence, for an IDB, means free from outside control and being capable of thinking and acting for oneself.

At first it might seem like a small difference from today's concepts of artificial intelligence, but it makes a massively different entity, and that entity has more extensive capabilities then today's systems using AI.

All human brains can be considered to operate as "Independent Analog Intelligence".

Here is an example showing the difference between today's type of AI and IDI:

- A company makes golf balls and Jack works for that company. Jack's job is to sit at the end of the production line and take four golf balls out of a bin, put them in a specific box, close the box, and put the box of golf balls on a conveyor belt that sends the boxes to the shipping department.
- The company is growing quickly and the manufacturing department requests adding a robot to do the same job as Jack is doing to double the output of their product. The request is approved, the robot is ordered and received and sent to the engineering department to be programmed.

- The robot is installed next to Jack, they work together and double the company's output of their product.
- The company continues to grow and needs even more volume of their production. They decide that the next change in manufacturing to address this growth is to begin putting eight balls in each box. They make the new boxes.
- One Friday evening Jack's boss comes to Jack with the following message.
- "Starting on Monday, I want you to begin putting eight balls in each of these new boxes. Stop using the boxes used for four balls. Any questions?"
- Jack answers "I got it and will use the new process starting on Monday." The boss looks at the robot then calls engineering to reprogram the robot to use the new process.

Now if that robot had the new IDI technology and had also been listening to the boss talking to Jack about the new process, the robot would also have said "I got it and will use the new process starting on Monday."

Defining the term "bodies"

As an example, let us compare the IDIOS and IDB to the equivalent parts of a human:

- The IDIOS would be equivalent to the human brain.
- The IDB would be the total human body.

Another example would be to compare the IDIOS and IDB to the equivalent parts in a cell phone.

- The IDI would be the cell phone's processor and operating system.
- The IDB would be equivalent to the complete cell phone including the case, screen, button, battery, SIM card, etc.

The different types of IDBs are huge. Everything that contains a computer today, could be an IDB in the future and use an IDIOS instead of today's operating systems.

A partial list of possible candidates for IDBs are airplanes, cars, trains, ships, satellites, space stations, homes, commercial buildings, dams, heavy equipment, traffic control, companions, doctors, surgeons, engineers, farmers, miners, teachers from kindergarten to college professors, manufacturing robot machines, cell phones, large commercial computers, entertainment devices, information sources, and so much more.

Defining the term "operating system"

The concept of IDI adds no new meaning to the well-established concept of an operating system. An operating system manages the hardware and software resources of a computer system. The IDIOS does basically the same things but in the IDIOS, there is no need for human user interfaces and there is no application software. Therefore, and obviously, the IDIOS does not need, nor does it have resources to support these two things.

IDI and IDBs in the Beginning

Today there are no IDBs with IDI.

So, let us start with the first IDB using IDI technology. There always has to be a first one. The complete massive system of IDBs using IDI technology that will exist in the future will all be traceable back to this first one.

For explanation purposes and for my examples, the first IDB using IDI technology will be a modified laptop computer.

So we will not have to be calling it the first IDB using IDI technology let us give it a name. The first IDB using IDI technology will be called "Adam".

Because an IDB is independent, there is no need for any human interfacing devices.

Take the keyboard off and replace it with solar cells so the IDB can charge its batteries independently.

Next, remove the power switch because the IDB will always be on.

Next, remove the touch pad as it is a human interphase device.

Next remove all applications (Apps) as they are intelligence created by human programmers.

Retain the CPU.

Replace the operating system with the IDIOS.

Maintain the needed "drivers" that make the video camera, the speakers, the microphone, and the monitor screen work properly.

This modified laptop will now represent the first IDB I will call Adam.

I will also take the liberty of calling Adam a "him" even though IDIs have no gender.

Turning on the IDI for the first time

Wait, there is no on/off switch on Adam. Move Adam to a table next to a window that gets lots of sunlight.

After the solar cells get enough sunlight, they will charge the internal batteries and the internal digital circuitry will come to life.

At this point in time, what are the capabilities Adam has even before starting to sense his environment?

The CPU and IDIOS will be "up and running".

The input and output devices will be active.

For Adam, those inputs and outputs would be hearing, seeing, making sounds, outputting to the monitor, and connecting to a communication feature that will allow an IDB to communicate with other IDBs and human networks. More about the networking of IDBs will be addressed later in this book.

Because this is the first time Adam has the CPU and operating system running, in human vernacular, Adam would now be "born".

Differences between a new IDB and a human baby

It takes a while for a human baby to get all its senses to the level of an adult human. Adam will be born with all senses at their highest level. Another way to think of this is there are no baby or adolescent IDBs. Just adults!

Both a newborn human baby and Adam will have their "brain" ready to learn and create knowledge upon birth.

Both will build their knowledge based on sensing their environment.

The human baby will change physically. Adam will not change physically.

Both the human baby and Adam will learn based on a similar process. That process will be explained in detail in this book.

IDBs will be unique and independent (like humans), therefore they need unique names.

Present day computers do not have unique names. If you are computer savvy, you might say what about a media access control address (MAC address) or a universally unique identifier (UUID)? Human users of a computer can change any of these "computer names". Therefore, our present computers do not have unique names independent of human interference.

The only way for each IDB to have a unique name is for a worldwide entity or company to assign names for every single IDIOS manufactured. That is not that difficult. In our society, we have Social Security numbers for ourselves and IP addresses for our computers' networking names. Even this book has a unique identification called an "ISBN".

Not only will each IDIOS have a unique name, but each memory location will have a unique name. That means that each one of Adam's memory locations will have a unique name in the complete world of all IDBs' memory locations.

In present computers a memory location has one function and that is to store a piece of information. In the IDIOS, each memory location does several other tasks. There is no equivalent element in present computers so we will need a new name to address these new computer entities. In my first book, I called them "memens" so I will also call them "memens" in this book.

You pronounce "memen" as you would the two words "me" and "men".

It may seem difficult to have a unique name for every single memen in the world, but it is quite easy. Simply attach the unique IDB's name to the front of the address of each long-term memory memen in an IDIOS.

I will explain the function of different types of memens in the following chapters.

How does the IDIOS store information?

First of all, what is long-term memory?

Long-term memory is where humans and IDBs store their information and knowledge. In an IDB it can remain in their memory forever. In humans, it will deteriorate with time.

For every sense the IDB contains, there is a sensor. So, for Adam, there is a visual input sensor which would be a video/picture device like a webcam. Adam would also have a sound sensor which would be a microphone. Each sensor would be connected to an input monitor memen. For every input sensor, there is one and only one input monitor memen. For Adam, there would be one sound input monitor memen and one visual input monitor memen.

We know we do not put into memory everything we sense. Humans and Adam sense many things per second with their eyes and ears. So how do we or Adam determine what is worth saving into memory for future use? It is quite complicated but a good English word to define what is important in what you sense and want to put into memory is the word “curiosity”. If a human’s or Adam’s curiosity reaches a high enough level about something being sensed, their input monitor memen will want to save it into long-term memory.

Curiosity will be based on characteristics of the sensing device. For example, with a microphone as an input device, curiosity could be based on sound intensity, sound familiarity, a continuous sound, or many other characteristics.

Every single long-term memory memen has several functions but one of the functions is to constantly listen to its input monitor memen.

All long-term memory memens are not the same. They are customized based on the information they store. For Adam, his visual long-term memory memens will store pictures and videos so these memens will store information in formats like jpeg or mpeg. His audio long-term memory memens will store sounds so these memens will store information in formats like mpeg, wave, mp4, dvi, etc. All long-term memory memens, some being empty and some containing information, will always be listening to their input monitor memen.

When an input monitor memen has been curious enough to want to store information into long-term memory, it calls to all long-term memory memens and asks, "Let me know if you are empty and if so, send me your address." Empty long-term memory memens will let their input monitor memen know they are empty and send their address. The information will be stored in one of the empty long-term memory memens.

Curiosity will decrease after the input memory memen stores information into long-term memory. Curiosity also decreases after the input memory memen learns what it is trying to save is already in long-term memory.

Once Adam is curious enough about something in his environment and stores it into long-term memory, it will last forever. We humans tend to forget some information or knowledge we learned if we do not pull it out of memory and use it. Do you remember everything you learned in all your English, history, and math classes? Of course not. Adam would have remembered everything.

Information in long-term memory lasts as long as the IDIOS remains functioning. If you remove power to the IDIOS, the information in long-term memory could be lost if it is not in permanent memory.

I have created several parts of an IDIOS and installed them on a laptop computer in my home. It looks out my back window with a high-definition webcam. What the IDB sees is constantly being updated on the laptop and displayed on its monitor. When it becomes curious enough to want to store a piece of information, it places a green dot over that information on the laptop monitor. The green dot will be moving around as it chooses to watch different things. Maybe a golfer then a bird then the golfer again then a flower moving in the wind or a person in the hot tub. It is quite mesmerizing!

What exactly is knowledge?

For our discussions, information or data is a single thing. Here are a few examples of information: a horse, the color brown, a word, a name, the act of skipping or running.

Knowledge is a set of related pieces of information. Here are a few examples of knowledge: a brown horse, The Grand Canyon, a Boeing 747, or a shod brown American Quarter Horse colt.

Some knowledge can have thousands or millions of pieces of related information. For example, if I ask you to tell me everything you know about horses, that list could be large. An example of someone's knowledge could be: four legs, mane, tail of hair, neighs, saddle, reins, runs fast, eats grass, colt, mare, palomino, roan, humans race them, can be shod, can be ridden by humans, and so much more. A Farrier would have a high degree of knowledge about a horse. Another example is if I asked you to tell me everything you know about the document called The Magna Carta of 1215. Your knowledge of this document could be small with maybe less than 10 pieces of information.

Some types of knowledge are a related collection of other pieces of knowledge. An example is mathematics. Mathematics has several fields of knowledge. Examples would be adding, subtracting, multiplying, geometry, trigonometry, calculus and much more.

In conclusion, we now know how Adam stores information in each of his long-term memory memens.

Sets of related information is called knowledge. That is what knowledge is but how does Adam turn information into knowledge?

How does the IDIOS create knowledge?

The key to the creation of knowledge is the activity that happens within short-term memory. I am sure you have heard of short-term memory in humans. I make no claim that IDI short-term memory is the same or even similar to short-term memory in humans. I will be only addressing IDI short-term memory.

Short-term memory is the only place within Adam where information can be stored for possible use in creating action or thought.

Short-term memory is composed of millions of short-term memory memens.

All information in short-term memory came from long-term memory.

Short-term memory memens have no unique address and assume the address of the information it gets from long-term memory.

Short-term memory memens and long-term memory memens have the capability to communicate with each other.

All short-term memory memens contain uniquely different information.

When a new piece of information is brought into short-term, it will contain an intensity level. New information going into short-term memory will have an intensity of one.

With one exception, the intensity of all short-term memory memens will always be declining. When the intensity declines to zero, that short-term memen clears its memory and is available for reuse.

A second function of all long-term memory memens that contain information is to constantly monitor the input to their input monitor memen. The goal of this activity is to find their piece of information in what is being sensed by their input monitor memen.

While a long-term memory memen is sampling the environment and recognizes its information in the environment and becomes curious enough, it will want to transfer a copy of its information to a short-term memory memen. It will ask all the memens in short-term memory "Do any of you contain my information I am about to send to you?". If nobody responds, the information will go into an empty short-term memory memen with intensity of one. If a sort-term memory memen responds and says, "I have your information", the long-term memory memen will add the new intensity of one to the existing intensity of that short-term memory memen.

That same long-term memory memen will also talk to all long-term memens with its related information addresses, and that causes all those long-term memory memens to also send their information to short-term memory.

For example, Adam goes to a farm and sees a horse for the first time, becomes curious enough, and stores a picture of a horse into a long-term memory memen.

Adam continues looking at the horse, and the long-term memory memen with a picture of a horse increases its curiosity and sends its information to a short-term memory. That short-term memory memen now has a picture of a horse with intensity of one.

Adam goes home for a day and the intensity of the picture of a horse in short-term memory is decreasing. The next day, Adam's short-term memory of a picture of a horse may have decreased to one half. He goes to the farm again and sees a horse again. The long-term memory memen containing a picture of a horse gets curious and sends its information to the short-term memory memen of a picture of a horse and the intensity increases to one and a half.

In conclusion, if a specific piece of information is in short-term memory and long-term memory sends that same information to short-term again, the intensities will add together.

Now, if you visualize what short-term memory would look like, it would be millions of informational memens with their intensities constantly changing based on what Adam is perceiving in his environment. Some will be increasing, some decreasing, some new, and some going away.

Now that we understand how short-term memory works in an IDB, I will show how knowledge is created from that short-term memory.

For Adam to create knowledge, there needs to be a process to gather all associated information about anything Adam knows.

It all happens in short-term memory. Here is an example to explain the process:

Adam continually looks at a horse. The picture of a horse would be in long-term memory and short-term memory and its intensity in short-term memory would continue to increase until you stopped perceiving the horse.

Let us say you, as a human trainer, want Adam to learn the sound humans use to identify a horse. That piece of information would be a sound clip (.mpeg) of a human saying “horse”.

While Adam is constantly looking at the horse, you as Adam’s trainer say the word “horse”. You continue saying “horse” as Adam continues looking at the horse. Adam’s curiosity about hearing the word “horse” several times in a row increases until Adam will store the sound of the word “horse” into a memen in long-term memory.

Adam continues to look at the horse and you continue to say the word “horse”. Adam’s curiosity continues to increase, and Adam places the sound of the word “horse” (that is in long-term memory) into short-term memory.

We know there will be many pieces of information in in short-term memory including the picture of a horse and the sound of the word "horse".

Because Adam is constantly looking at a horse and you are constantly saying the word "horse", the two short-term memens containing the picture of a horse and the sound of the word "horse" will constantly increase in intensity

When the intensity is sufficiently higher than the other memens, the "magic action" happens.

The short-term memory memen containing the picture of a horse tells the long-term memory memen of the sound of the word "horse" to save the address of the picture of a horse in its memen.

Also, the short-term memory memen containing the sound of the word "horse" tells the long-term memory memen of the picture of a horse to save the address of the sound of the word "horse" in its memen.

Simply said, they exchange information.

At this point Adam can create knowledge with the help of a human trainer.

Adam's knowledge is the set of all long-term memory memens and their associated addresses of other long-term memory memens.

There is a form of information we have not yet addressed. It is what we call verbs.

So far, we have generally talked about knowledge being created with things (nouns), adjectives (things that create a distinction for a noun), and adverbs (things that create a distinction for a verb or another adverb).

Examples are:

- A dark brown horse.
- A large white house.
- A very funny clown.

The meaning of nouns, adjectives, and adverbs can generally be captured in a picture.

Of course, there is another type of word that cannot be defined by a picture. These words are called verbs. These words require a series of pictures over time to define the word.

Some examples are:

- Standing
- Walking
- Running
- Jumping
- Hopping
- Skipping
- Falling
- Flying
- Speaking

So, how does an IDB store a verb in long-term and short-term memory?

It simply stores a series of pictures in a long-term memory memen.

For example:

"Standing" could be portrayed by a series of pictures of a human being erect and with both feet on the ground. It could not be just one picture because that would depict a human as a noun. Maybe it would require sixty pictures to define standing.

"Walking" could be a series of pictures of a human moving forward by putting one foot in front of the other while maintaining at least one foot on the ground at all times. Maybe it would require sixty or more pictures to define walking.

"Running" could be portrayed as a human walking so quickly that at some times no feet are on the ground. Maybe it would require sixty or more pictures to define running.

"Jumping" could be portrayed as a human standing and pushing upward off the surface so that both feet are in the air. Maybe it would require sixty or more pictures to define jumping.

"Hopping" could be portrayed as moving forward by jumping. Maybe it would require sixty or more pictures to define hopping.

"Skipping" could be portrayed as walking or running intermixed with hopping. Maybe it would require sixty or more pictures to define skipping.

We are not done yet with verbs.

How would we store information in a long-term memory memen to portray Usain Bolt (the fastest man on Earth) running? Humans can discern up to sixty pictures a second. It would only take a few seconds to describe him running and that would be maybe two hundred pictures. Technology to show this many pictures at 60 pictures per second is with files we call .mpegs.

We still are not done with verbs.

How would we store in a long-term memen a dance routine like country western line dancing?

Line dancing is when dancers line up in a row without partners and follow a choreographed pattern of steps to music.

First the IDB would need to learn the steps. Each step would be a series of verbs.

Each step would be stored into own long-term memory memen.

Any dance would be a particular series of steps. Therefore, a dance would be stored into a long-term memory memen as a series of steps found in other memory memens.

We still are not done with verbs.

A IDB could listen to a song and store the whole song into a long-term memory memen. If it wanted to play the song, the IDB would transfer the song to a short-term memen and then play the sound on its speakers.

We still are not done with verbs.

How would an IDB dance to a song in a freestyle fashion?

When you go out onto the dance floor to dance, you do not have the complete dance planned out in your mind. You have learned "steps" and "moves", and they are part of your knowledge of how to dance. You will use these moves during the dance, but you have no idea what you will be doing exactly in the next few seconds. How does this work since you are confident you will complete the dance with no problems?

It is all about the feedback loop between what you are doing and how you perceive, through your input sensors, what you are doing. All of your moves are in short-term memory and their intensities are constantly changing based on your curiosity about what you are actually doing. The moves with the highest intensity in short-term memory are going to be your next moves. That is how it will work for IDBs.

This constant feedback loop, for humans and IDBs, is operating 24 hours a day 7 days a week. It is imperative for everything we and IDBs do.

How do IDBs share their information and knowledge?

We first must address how IDI's communicate wirelessly or remotely.

No matter how humans communicate, there are always problems to get the intended information and knowledge transferred between two humans.

The simplest example to show this fault in humans is the age-old example when ten people sequentially whisper to the next person a specific statement. After the whispering is done, the first person reveals what he said, and the last person reveals what he heard. They usually are quite different.

If IDBs did this same example, and communicated wirelessly to each other instead of whispering, the specific statement received by the last IDB would be exactly as it was from the first IDB 100% of the time.

A second example to show a fault in human communication is when one person with a particular piece of knowledge wants to communicate that knowledge to another human. After the explanation is over, the first person is not totally sure the second person "gets it" and the second person also is not sure he "got it".

Here is an example we can use between you and me. I will use in the example a particular animal you probably have never heard of or seen. After my explanation of the animal, try drawing it.

After drawing the animal, Google the animal and see how well you and I did.

- The animal is a Glaucus Atlanticus.
- It is small and lives in the ocean. It is only an inch or two long. It is a slug like animal that is actually a mollusk. It is silvery grey on its bottom side and dark and pale blue on its top. It has dark blue stripes on its head. It has a flat, tapering body and six appendages that branch out into rayed, finger-like appendages.
- Now draw what you think it might look like.

How well did we do?

If one IDB, with the knowledge of a Glaucus Atlanticus, communicated its knowledge with another IDB wirelessly, the knowledge in both IDBs would be identical.

How do IDBs exchange knowledge remotely?

First, humans would have to design a different wireless protocol than we have today. Todays wired and wireless protocols are information-based or data-based. For IDBs, it would have to be knowledge-based.

Second, each IDB would have to have capabilities of connecting to the internet.

Once IDBs have this capability, they can communicate with each other.

It actually is quite simple. They will be sending knowledge, not data, so the smallest piece of communication will be the information along with all the addresses of the information's associated data.

There needs to be two different entities to explain how this communication would work. Adam is one of our entities, but we need another. I will call the second IDB "Eve". I find it easier to understand when you use human names instead of IDB "A" and IDB "B".

Remember that each long-term memory memen address is unique among all long-term memory memen addresses in all present and future IDBs.

There are several different possible combinations that need to be implemented. Here they are presented one at a time.

The first case is with Eve giving all her knowledge to Adam over the internet.

For this example, she will send Adam all her knowledge in all her long-term memory memens.

First the communication link is established.

Information comes across the communication link in packets.

Each packet consists of the information in one long-term memory memen followed by all the memen addresses that are associated with that memen. Remember that if Eve is deeply knowledgeable about something, there could be thousands or even millions of associated addresses.

Adam receives the first long-term memory memen from Eve.

Adam checks to see if he has a long-term memory memen with that address.

Let us assume Adam did have this memen's address in the information Eve transferred to him.

What would that mean? It would mean that in a previous communication Adam shared that piece of information with Eve. Adam would know that because the address of that piece of information would start with his IDI number. Adam would therefore not put that memen into his memory because he would then have two long-term memory memens with the same address. That is not allowed.

But we are not done. Maybe after Adam sent his knowledge to Eve, Eve learned more and added more knowledge in her memory. Adam would want that new knowledge Eve had gained.

To accomplish this, Adam would have to look at each of the associated addresses and see if he had that information. If Adam had the information, he would skip to the next address. If Adam did not have the information in the next address, he would create a new long-term memory memen with that address.

At this time, the new long-term memory memen would contain no primary piece of information. At some time during the transfer of information, Eve will send the information in a long-term memory memen that contains the information Adam needs for his empty memen. That information will be transferred into Adam's empty long-term memory memen.

As Adam continues to get information from Eve, any associated information with Adam's new long-term memory memen would be discovered and added as a piece of associated information.

Let us assume Adam did not have the memen's address Eve sent to him.

Adam would simply add it to his long-term memory by creating a new long-term memory memen and continue the transfer.

After Eve shared all her long-term memory, Adam's long-term memory would contain an exact copy of Eve's long-term memory.

Adam may have some knowledge that Eve does not have. If we used the same process with Adam sharing his knowledge with Eve, then their long-term memories would be identical. All pictures of one IDB would be an exact copy of the pictures in the other IDB. It would be the same of all sounds, smells, touch, and other possible IDB inputs.

Humans teach IDBs differently about the same thing

Let say we have 10 IDBs that all have a long-term memory memen with the picture of a horse. In reality, if they had not shared information, they would have many pictures of a horse, all different and all with different addresses. Let us say each of the 10 IDBs was trained by ten different humans who each spoke a different language. Once they all shared their knowledge, they would all have ten different human sound voices in ten different languages of a horse in their long-term memory. A similar situation happens with humans learning in different places with different languages. It becomes more complex for IDBs because they do not know when or where they got their knowledge.

This problem exists only with words learned from humans. IDBs can solve this problem by storing the language of the human word as a piece of associated data.

How does the IDIOS output to its IDB?

The source for all IDIOS outputs comes from short-term memory.

That makes sense because the relationship between output memens and short term memory memens is very similar to the relationship between input memens and long term memory memens.

What are some types of IDB outputs?

There is a sound output memen, a display output memen, and many kinematics output memens, and wired and wireless communication memens, and more.

Kinematics is simply the branch of mechanics concerned with the motion of objects.

There is one output memen for each output function. Each output memen has a unique address. Each long-term memory memen that could possibly cause an output or action (in other words a memen that contains a "verb") contains the address of the output memen that would cause the action.

For example:

Long-term memory memens containing words like "getting", "talking", "running", "looking", or "touching" all have at least one address of an output memen that could accomplish the activity.

Long-term memory memens containing a horse, a color, a smell, the word "big", or the word "high" all have no addresses associated with an output memen.

The short-term memory memens with the highest level of intensity are the candidates for output. When these intensities are high enough, it will trigger what I will call the activity of "motivation".

Motivation is simply the activity that causes something to happen. Therefore, high enough levels of intensity in one or more short-term memory memens will always cause something to happen.

The actual level that would cause an output is constantly changing. As the level of different short-term memory memens intensity increases, so does the level needed to cause an output. This is similar to the function of curiosity on the input side of the IDB.

Here is a simple example of the process to create an output:

Almost all outputs require not only outputting to a device but also considering input information coming into the IDB.

For example, Eve might ask Adam to neigh like a horse. Adam could do that if he had ever heard a horse neigh. If Adam did not know how to neigh, he might ask another IDB that resides on a horse ranch to send him all his information about a horse. That information would include the sound of a horse neighing. Adam could then create the horse sound for Eve.

A second example is a request from Eve for Adam to touch his nose.

First the intensities in short-term memory need to be sufficient to initiate the touching process.

There are several things that are required to accomplish this action and they need to have been learned before this process will be successful. The process goes something like this:

How is Adam going to touch his nose? He chooses to use the tip of his index finger on his right hand.

Then there is the process of moving his finger from its present position to a position where the tip of the finger is touching his nose. He has learned the closed loop process of moving his finger. That includes visually monitoring the tip of his finger position and moving the tip of his finger again to be closer to the final goal of touching his nose. A key point here is that no two humans will use the same path for the tip of the finger to accomplish this task. That is also true for IDBs. These movements are learned processes for each human or IDB because they are physically all different. Today's assembly robots are programmed by humans and a programmed move is the same every time unless programmed otherwise.

There is also the dynamics of touch, or pressure. If Adam touches his nose, then Adam would feel a pressure.

Of course, it would be important to tell Adam what pressure to use. Over time, Adam may have a default pressure when touching himself. Humans certainly have a default pressure when touching themselves. In this case, it would be assumed that "touching the nose" is very different than "smashing the nose".

Kinematics is dependent on the physical structure of the IDB. Let us say that Eve has learned to play the piano. She could share her knowledge of how she plays the piano with Adam but if Adam did not have the same appendages that are required to play a piano as Eve, he would not be able to play the piano based on the information received from Eve.

A quick overview of each element of an IDB

An IDB is composed of five main elements. Here are the different elements along with their type or function:

1. **Input Sensor Types**

- Vision Sensors
- Sound sensors
- Motion sensors
- Wired or wireless electronic communication
- Temperature sensors
- Pressure sensors
- Modems
- Network cards
- Touch Screen
- Moisture/humidity sensors
- Magnetic field sensors
- Radar sensors
- Gas sensors
- Light sensors
- pH sensors
- etc.

2. Input Memen Functions

- Constantly monitor the environment.
- Determine if curiosity is high enough to save what is being sensed.
- Request from long-term memory an empty long-term memory memen along with its address.

- Listen for the response from all the long-term memory memens.
- Once the input memen receives the address of an empty long-term memory memen, it stores what is being sensed into that empty long-term memory memen.
- It has the ability to change its level of curiosity.

3. Long-term Memory Memens' Functions

The function of an empty long-term memen:

- Listen for the input memen to request an empty long-term memory memen's address.
- Respond to the input memen with your address.
- Accept the information from the input memen.

The function of a long-term memory memen that contains information:

- Constantly scan the environment is see if your piece of information appears in that environment.
- If it does appear, send it to short-term memory.
- Notify all the associated long-term memory memens whose addresses you have as related information. They will also send their information to short-term memory.

4. Short-Term Memory Memens' Functions

- Short-term memory is constantly listening to all long-term memory. If a long-term memory memen requests to send its information to short-term memory, one of two things will happen:

  - If the information from the long-term memory memen is not already in short-term memory, short-term memory adds it to an empty short-term memory memen and assigns the short-term memory memen with an initial intensity level.
  - If the information from the long-term memory memen is already in short-term memory, short-term memory adds an initial intensity level to the existing intensity level of the short-term memory memen.

- Once an intensity level of a short-term memory memens is established it will constantly be decreasing. When its intensity gets close enough to zero intensity, it will delete its informational contents.
- If the intensity of any short-term memen is relatively higher than the other memens, it can send its information to the associated output memory memen for implementation.

5. Output Memens Function

- There is one output memen for every output device.
- It is constantly listening to all short-term memory memens.
- If it hears its address, it outputs to its output device for implementation of the action.

6. Some Output Devices

- Monitors
- Printers
- Plotters
- Projector
- Speakers
- Modems
- Network cards
- Touch Screen
- All kinematic types

The output device and its associated driver allow the IDB to interact in its environment.

How creativity works in humans

This subject is one that many find to be the most interesting.

We know and have experienced people who are creative in their field and others that are not so creative. We have our own perception of our own level of creativity, yet it seems hard to quantify and understand.

I will address creativity in IDBs and how it works exactly but I want to introduce the subject with a short story.

As a manager in a large and well-respected company where technology was king and a significant key to the company's success, I had new engineers and SEED (Student Employment Experience Development) students asking the question about how they could learn to excel in creativity. The reason for their question is because they knew, to grow in their engineering field, they need to have patents or trade secrets in their name either as the primary inventor or as part of a team of inventors. Of course, in a large company, these patents are owned by the company. Being an inventive engineer is a great skill and will work in their favor when applying to be a technical lead or technical fellow.

The story goes like this:

A young engineer asks how to increase their level of creativity. I would ask the young engineer or SEED student if their name was on a patent or if they knew anyone that had their name associated with a patent. Generally, the answer was "no", so I gave them the names of a few engineers within the division that had their names on patents and told them to talk to these people. I asked them to ask the engineers "How or why are you creative and others are not"?

We would meet again within a week, and I would ask them what they had learned.

The answer was always the same. The inventors will say they really are not sure why it was them and not others. They will also make a very important point and that is that other engineers had the same knowledge they had which was needed to make the invention.

Another point that would be made is that all the knowledge required to make the invention was well-defined. In other words, there is no unknown "magic" that was required.

This would generally be a good enough understanding for the young engineer to charge ahead feeling they have as good a chance as anyone else to succeed at getting their name on one or more patent in the future.

The key point to this story is that many engineers have all the knowledge needed to make the discovery.

Potentially by humans understanding how IDBs act creatively, we could be more creative.

How creativity works in the IDIOS

If there is an opportunity to be creative within an IDIOS, all the knowledge needed for the creation needs to be in long-term memory.

Also, knowledge must be put in short-term memory in order for the IDIOS to act upon the knowledge and be creative.

Here is a short story to demonstrate creativity:

Your daughter comes home in her car and parks on the street in front of your house.

In the process of her getting out of the car and walking behind her car while putting her car FOB in her purse, she accidently dropped the FOB and it fell into the water drain which is covered by a grate. It drops about fifteen feet to the bottom.

Your daughter asks you to help get her FOB back.

After you go outside and see the situation, you begin thinking of a way to retrieve the FOB. The FOB is mainly plastic with no ring for other keys and a smooth surface.

You bring in knowledge to implement the act of "getting" the FOB. None of your knowledge associated with these actions can retrieve the FOB. You thought of hooks, magnets, loops, and pinchers but no success. The intensity of the "getting" functions is high in your short-term memory but not sufficient to complete the task.

You need to get creative but how?

You go in the house to think more about the solution.

You walk upstairs and walk by your 8-year-old daughter playing in her bedroom on the floor with one of her large dolls.

The doll seems to be standing on the floor by itself.

Your curiosity is enough to go check it out.

Your daughter had put a blob of Silly Putty on the floor and then put her doll's feet into the Silly Putty until it formed around the doll's feet and supported the doll to stand.

You know the rest of the story because it just happened to you!

Dad takes some Silly Putty outside, ties a piece of string around it and lowers it onto the FOB. After just a few minutes the Silly Putty forms around the FOB and Dad pulls the piece of Silly Putty along with the FOB out of the drain.

That is the story but what happened? I will explain what happened if Dad were an IDB.

Dad moved his long-term memory memen that contained the verb "get" to short-term memory and it has a high intensity. Related information (or knowledge) in the "get" long-term memory were "by a hook", "by a magnet", "by a pinching device", "by a scooping device", "by a blowing device", "by gluing", "by digging". The whole concept of "getting" was very high in Dad's short-term memory.

Under a separate situation Dad observed Silly Putty "getting" his daughters doll to stand by surrounding and holding the doll's feet.

Therefore, Silly Putty and a piece of knowledge of Silly Putty which is its ability to "get" go to short-term memory.

The "get" memen from Silly Putty added to the "get" memen from Dad attempting to find a "get" solution for getting the FOB. This causes short-term memory to add Silly Putty as a piece of knowledge for the "get" long-term memory memen.

Dad now has knowledge to get the FOB with Silly Putty.

Another way to look at creativity in IDBs is as follows:

When two pieces of unrelated knowledge (knowledge "A" and knowledge "B") are introduced to short-term memory and one of knowledge "A's" piece of related information is the same as a piece of related information of knowledge "B", the related pieces of information add their intensities. If the related pieces of information intensity is high enough, they share that information and send that new information back to long-term memory in the memens containing knowledge “A” and “B”.

In conclusion, there are two separate functions that populate long-term memory. The first is curiosity about the environment which creates knowledge. The second is what humans call creativity which is the process of short-term memory identifying a trait that is common to two pieces of otherwise different knowledge in short-term memory and sharing it back to long-term memory.

Clarification of the meaning of "curiosity"

As mentioned earlier, an IDB's curiosity is just a number that is constantly changing in each of its input memens. Different input memens have different functions that cause their curiosity to change.

This might not seem intuitive but the lower level of activity of a specific input produces the highest level of curiosity. For example, if Adam were in a room with total darkness and the smallest "pinhole' of light appeared somewhere in the room, it would trigger curiosity. If an IDB was entering a stadium with 100s of other people or IDBs crowded together, curiosity of the general mass of people would be quite low.

As in humans, increased curiosity results in increased amounts of information being stored in long-term memory which results in more knowledge being formed in short-term memory.

In, short, curiosity is the basis for knowledge in humans and IDBs.

There is an advanced form of curiosity once a form of communication is established between IDBs or IDBs and humans.

One example is when one says to another "Do you see the rabbit?". A second example is when one says to another "Watch out!". I believe there has been enough information presented for you to define how this would work.

Clarification of the meaning of "motivation"

One of the primary feedbacks I get about the IDIOS is "I understand the process but where is the software to control the process and emulate the human mind?"

My answer is as follows:

Let us assume we need to have a programmer program more software to control the process and emulate the human mind.

So, a programmer programs what he considers to be that software.

I then ask the programmer if the process with the new software is independent from human intervention.

The programmer answers "Of course not. It does what I programmed it to do".

I then ask, "If that is the case, then is the old process plus your new code you wrote now the new process?"

The programmer's answer is "Yes. Of course.".

I then ask, "Do we need some new software to control the new process?".

The programmer's answer is "Well, yes.".

I then ask, "Would you write some new software that will act like the human mind that will control the software you already wrote to be the human mind that was written to control the original process?"

You can see this is an unending process.

The real answer is that Independent Digital Intelligence contains all the elements in its process to mimic the human mind.

This concludes the technical discussion of how an IDB works.

As I mentioned earlier, this explanation is not intended to present enough information for an R and D department to move the development of an IDIOS from the investigation stage to the lab stage where the software to create an IDIOS would be programmed.

This discussion with more technical enhancements would be sufficient to be called a feasibility report.

I hope it does present enough information for you to see the feasibility of humans creating an IDIOS for an IDB.

All the information and technology to create an IDIOS does exist today.

It will be a fun and exciting world with our new partners.

IDB's Contributions to Human Societies

Learning and Teaching

How an IDB will learn and teach was partially addressed in the Technical Discussion of this book. I talked about how Independent Digital Intelligence was designed to allow an IDB to be able to learn and teach. Here I will get more specific about how this will happen and its contributions to our society.

IDBs learning from Humans

An IDB, once it is built by humans and "turned on", has the capability to learn immediately based on sensing its environment. This is a great capability but not sufficient to learn the knowledge contained within the human race. Here are the pieces of knowledge that could take IDBs weeks, decades, or centuries to learn without the help from human teachers:

- Learning a language to communicate with other IDBs or humans both written and spoken in the various languages.
- Learning a numbering system to quantify the physical world.
- Learning the branches of mathematics that humans understand. That would include arithmetic which is the branch of mathematics dealing with the properties and manipulation of numbers, trigonometry, geometry, algebra, statistics, calculus, and more. One can think of these branches of mathematics as the language to describe the physical world.
- Learning physics.
- Learning chemistry.
- Learning medicine and medical techniques.
- Once these pieces of knowledge are taught, then humans can teach IDBs how to design and build all the human inventions. These pieces of knowledge include but are not limited to:

- Farming and mining.
- Making metals like bronze, iron, and steel.
- Making objects that create and distribute electricity.
- Making objects that manipulate electricity. Things like resistors, capacitors, inductors, etc.
- Making transistors.
- Making integrated circuits (also known as microchips). Integrated circuits are one piece of semiconductor material loaded with transistors and other electronic components.
- Making computers.
- Making objects that physically move other objects.
- Making vehicles like planes, trains, automobiles, trucks, and objects that can travel in outer space.

Once we teach IDBs these pieces of knowledge, the IDB community will have and be able to share this information with both existing and new IDBs forever. This is unlike the human community that must teach a new set of individuals each of these pieces of knowledge over and over again because what is taught to a human goes away when the human dies.

IDBs learning from each other

With IDBs, you just need to teach one IDB a particular piece of knowledge and it can be shared with all other IDBs as wanted or needed. There is another capability that IDBs will have that humans will never have. That capability will be the ability for IDBs to gain mental knowledge and physical knowledge without the IDB ever experiencing it. I consider this capability as being the third most important feature of IDBs. The first two are IDB "curiosity" and IDB "motivation" as presented in the first part of this book.

This third capability will be designed and built into IDBs by humans. It involves a wireless communication link between IDBs. Of course, this link would have all the features of human-to-human wireless communication. It will also have the IDB's unique capability to transfer knowledge to another IDB directly, wirelessly, and without either of the IDBs operating the transfer process.

How does this happen because humans generally believe they cannot flip a switch (maybe with a few parameters) in their mind and then have a whole process happen without their knowledge?

The answer is simple. Humans will build the capability into the basic structure of the IDB. They can open a channel with any other IDB (that has the capability) and communicate. If they want to share some piece of knowledge or even share all their knowledge, they can instigate a process to do so, and it will happen. After the process is completed, the IDBs will have new knowledge and the IDB will not be able to distinguish the new knowledge from the IDB's previous knowledge.

It should not seem all that strange to humans because we do something similar. The difference is that we must learn the process before we can execute it whereas the IDB never had to learn the communication process before executing it.

Another way to look at it is that IDBs can gain the equivalent of personal experience from other IDBs who actually had the personal experience. An IDB would not be able to tell the difference between its personal experience of its environment and that experience given to it by another IDB.

How IDBs gain knowledge without experiencing it

Let us look at how humans gain knowledge. When humans are born, they have minimal knowledge. It takes about 20 years before each newborn human becomes independent and can contribute to society. Then after an additional 50 years, that human ceases to exist and the 20 years of training is lost. That means a human society must relearn everything it knows every 70 years. It also means that about one third of the human society is always learning and not producing.

Let us now look at how IDBs gain knowledge. All IDBs' knowledge whether it is learned by the IDB or transferred to the IDB by another IDB is in a human understandable and useable format. Humans built the IDB, so they have access to all IDBs' long-term memory.

Let us assume an IDB is an old version or is somehow depleted or deficient. Let us also assume either humans or other IDBs want to upgrade this old version. How this would be done is to obtain the new version of the IDB, obtain the knowledge in the old version of the IDB, transfer the knowledge of the old version to the new version, and bring the new version to life. Humans or IDBs could do this continually as needed.

The essence of this process would be that any IDB could always exist and, in the human vernacular, any IDB would never die.

Given this learning process, once any IDB learned a particular piece of knowledge, that piece of knowledge could be shared with every single other IDB on Earth and beyond.

Do IDBs have emotions?

Words like "emotions" have different meaning to different people. In order to not get emotional in defining "emotional", I prefer to use different words to capture the human body's activity that causes these emotions. I call that activity "juice squirting". By juice squirting I mean the human body's activity to increase the levels of juices that create the sensations we call emotions.

Some of these juices we call adrenaline, testosterone, oxytocin, dopamine and norepinephrine, serotonin, and vasopressin.

All emotions are caused by juice squirting but juice squirting also causes other activities that require little or no conscious activity. These are things like normal heartbeat.

I only want to address the juice squirting that causes what we call emotions.

Here are examples of words we use to denote types of our emotions:

Acceptance, admiration, adoration, affection, afraid, agitation, agony, aggressive, alarm, alienation, amazement, ambivalence, amusement, anger, annoyed, anticipating, anxious, apathy, apprehension, arrogant, assertive, astonished, attentiveness, attraction, aversion, awe, baffled, bewildered, bitter, bitter sweetness, bliss, bored, brazen, brooding, calm, carefree, careless, caring, cheerfulness, claustrophobic, coercive, comfortable, confident, confusion, contempt, content, courage, cowardly, curiosity, cynicism, dazed, dejection, delighted, demoralized, depressed, desire, despair, determined, disappointment, disbelief, discomfort, discontentment, disgruntled, disgust, disheartened, dislike, dismay, disoriented, displeasure, distraction, distress, disturbed, doubt, dread, driven, dumbstruck, eagerness, ecstasy, elation, embarrassment, empathy, enchanted, enjoyment, enlightened, enthusiasm, envy, euphoria, exasperated, excitement, expectancy, fascination, fear, focused, fondness, friendliness, fright, frustrated, flee, gloomy, gratitude, greed, grief, guilt, happiness, hate, helpless, homesickness, hope, hopeless, horrified, hospitable, humiliation, hurt, hysteria, idleness, impatient, indifference, indignant, infatuation, infuriated, insecurity, insightful, insulted, interest, intrigued, irritated, isolated, jealousy, joviality, joy, jubilation, kind, lazy, loathing, lonely, longing, love, lust, mad, melancholy, miserable, mixed up, modesty, moody, mortified, mystified, nasty, nauseated, negative, neglect, nervous, nostalgic, numb, obstinate, offended, optimistic, outrage, overwhelmed, panicked, paranoid, passion, patience, pensiveness, perplexed, persevering, pessimism, pity, pleased, pleasure, politeness, positive, possessive, powerless, pride, puzzled, rage, rattled, regret, rejected, relaxed, relieved, reluctant, remorse, resentment, resignation, restlessness, revulsion, ruthless, sadness, satisfaction, scared, scorn, self-confident, self-conscious, self-critical, self-loathing, self-motivated, self-pity, self-respecting, self-understanding, sentimentality, serenity, shame, shameless, shocked, smug, sorrow, spite, stressed, strong, stubborn, submissive, suffering, sullenness, surprise, suspense, suspicious, sympathy, tenderness, tension, terror, thankfulness, thrilled, tired, tolerance, torment, triumphant, troubled, trust, uncertainty, uneasiness, unhappy, unsettled, unsure, upset, vengeful, vicious, vigilance, vulnerable, weak, worried, worthy, and wrath.

I would guess you had juice squirting happening based on the length of this list!

The reason for the long list is to explain that an IDB has none of these characteristics because they do not have a juice squirting function.

I believe IDB's lack of emotions will be a positive trait for our coexistence.

Even with no juice squirting characteristics, IDBs could learn the meaning of these emotion-based words and use them. Certainly, humans would communicate to IDBs expressing their emotions.

For example, politicians, movie stars, and music stars quite often tell their audiences they love them. IDBs could learn from humans how to use words like "love" in this type of context. It is not a juice squirting experience for the IDB but a way to effectively communicate to humans.

Without emotions why would an IDB do anything?

Humans without emotions, like curiosity and motivation, would probably do very little if anything at all.

With IDBs, it is different. As part of the IDB's operating system, there are two things that are built into the IDIOS that require the IDB to constantly execute. I detailed these two things in the technical discussion of this book, and they are what I call "curiosity" and "motivation".

Curiosity is just a number that is constantly going up or down based on what is being inputted to any one of the input memens. Motivation is just a number that is constantly going up or down based on what is in short-term memory and will result in the output memens performing an output. The IDB does not control either of these two operations.

Therefore, the IDB has no choice but to accept inputs that will constantly create the IDB's knowledge and constantly create output for the IDB to perform.

Humans could control both of these activities if they chose to do so.

If a human set an IDB's curiosity number to zero, the IDB would stop adding any inputs to long-term memory causing the IDB to stop learning. A human could set the threshold for curiosity higher and the IDB would store more information into long-term memory. This would accelerate the amount of knowledge that the IDB would gain.

If a human set an IDB's motivation number to zero, the IDB would stop all output from the IDB. A human could set the threshold for motivation lower and the IDB would output more activities. This would create what humans might call a "hyper" IDB.

If a human were to reduce both numbers to zero, the IDB would be in a deep sleep. Activity would still be happening in short-term memory which could be considered dreaming.

Characteristics of our new IDBs

From the human point of view, IDBs can be categorized into three different groups:

IDBs as tools for both humans and other IDBs

These IDBs could be called IDB tools. Here are examples of IDBs operating as tools:

- Trains, planes, and automobiles being IDBs for transportation.
- Our homes and our social and work buildings for shelter, protection, personal space, and workspace.
- Our personal and business computers performing their multitude of tasks.
- All of our energy sources like dams, windmills, solar grids, nuclear plants, fossil fuel plants, etc.
- All of our telecommunication network and plants.
- Financial institutions.
- Social media companies.
- Traffic control.
- Performing all our manufacturing functions in our manufacturing plants.
- Etc.

Yes, they will be the supporting structures in every aspect of our life.

IDBs who will make things

These IDBs will be called IDB builders. Here are examples of IDBs making things:

- Building our infrastructure. This is the underlying structure of a country and its economy, the fixed installations that it needs in order to function. These include roads, bridges, dams, the water and sewer systems, railways and subways, airports, and harbors.
- Building trains, planes, automobiles, and space craft.
- Building tools.
- Building homes and other buildings.
- Etc.

IDBs who will do things

These IDBs will be called IDB doers. Here are examples of IDBs doing things:

- Design things.
- Personal support for humans.
- Shopping.
- Cooking.
- Cleaning.
- Entertaining.
- Transporting.
- Teaching.
- Actuarial functions.
- Etc.

Because of IDB's ability to be diverse, versatile, scalable, and intelligent, they will support the foundation for our society.

Needs to sustain an IDB who is a doer or builder

I will address these needs by contrasting IDBs and humans needs.

- Does an IDB need a form of shelter or "home"?
  - Yes. IDBs are made from electronic parts that can be compromised by the elements. The size of an IDB's home could be just a little larger than the IDB.
- Does an IDB require a kitchen in the home?
  - No.
- Does an IDB require a bedroom in the home?
  - No.
- Does an IDB require a bathroom in the home?
  - No.
- Does an IDB require closet or storage space in the home?
  - No.
- Does an IDB require a security system to protect valuable assets in the home?
  - No.
- Does an IDB require electricity in the home?
  - Yes. Electricity is the food for IDBs.
- Does an IDB require a form of electronic communication?
  - Yes. In most cases IDBs would require a form of wireless communication for access

to other IDBs, humans, and other resources.

In conclusion, all that is required to support an IDB is an electrical connection, wireless communication, and a shelter just a little larger than the IDB. That is it!

Other interesting characteristics of IDBs

Another very interesting characteristic of IDBs is they don't have to look or act the same. They will look different and act different based on their function.

Another very interesting characteristic of IDBs would be their vocal ability. Of course, it is very different than our vocal system. We have lungs, vocal cords, an oral cavity, a nasal cavity, a tongue, and more. IDBs have speakers. IDBs can place sound files in short-term memory, then amplify and play the sound through a speaker. That could be a series of words, but it also could be a complete song.

IDBs will have to associate not only words with things but also the language of the words. IDBs would need to be multilingual if they received knowledge remotely.

IDBs over time might find it useful to adopt either an existing language or make their own language for their electronic transfer of knowledge.

IDBs could learn to play our musical instruments in a very different way. Humans dedicate years to learn the physical art of playing an instrument. Fingers moving on a keyboard. Fingers pressing buttons on a wind instrument. Fingers strumming on strings and pressing on frets.

Here is how an IDB might learn to play the piano:

The IDB listens to the sounds of each key on the keyboard under different pressures and timing and stores each sound into long-term memory. Next the IDB reads a piece of music and, instead of using fingers on a keyboard, it retrieves the sound of the notes from memory and plays them through its speakers. This could be done for any instrument including drums, guitars, saxophones and more. Potentially an IDB could play and mix many instruments in real time and truly be a "one IDB band"!

I have presented the idea that long-term memory lasts indefinitely in an IDB but how do you change something that was learned wrong? For example, a child growing up in the city, learns the word "dog" to associate with the family dog. The child gets praise by pointing at the dog and saying "dog". The child goes on a family vacation into the farm country. While driving down the road, the child sees a cow and points and says "dog". Nobody says anything and the child stores the picture of a cow in long-term memory with a piece of associated information being the name of the cow as "dog". Now the picture (or video clip) of a cow has associated knowledge of the word "dog" as an English language word that is associated with a cow.

Sometime in the future the child correctly learns the word "cow" as the English word associated with a cow. Now the child has the word "dog" and the word "cow" as words to represent a cow. We know that items in our long-term memory stay for a long time and we do not have a memory erase capability so how does the child know which word to use?

OK, let us analyze this situation.

The child had seen a dog and stored the images of dogs in long-term memory. Probably there was a dog in the child's home.

Someone had taught the child that the English sound that is associated with a dog is the sound "dog".

The child had used the sound "dog" when around a dog and the child was praised for using the right sound to identify a dog.

The child saw a cow and stored it in long-term memory. Since the child thought it was a dog, the child said "dog". Now a picture of a cow and all the knowledge of a dog are in short-term memory. The knowledge creation process, previously explained, happens and the dog sound is a part of the knowledge of the cow image.

The child, at a later time, sees a cow or even a picture of a cow and proudly says "dog" to the parents.

The parents tell the child it is a cow and use the sound cow till the child says "cow" and gets praise.

The child learned two different things. The child learned the sound "cow" to be associated with the picture of a cow but also a verb. That verb was "this is a cow so use the sound "cow" to identify it". This is a very common human process, and it is the very same process as used to learn a new language. All IDBs will need to be bilingual so this process will be used continually.

IDBs need to understand time

IDBs will need to use a time system to position an event in time. It will be important for IDBs to communicate timeframes to humans therefore they must understand the time that we use for our watches and clocks which is Ordinary Civil Time.

Here is an example:

3:34 Thursday, September 9th, 2021, at 3:34 PST

Ordinary Civil Time can become complicated. When you travel from place to place, you will need to change the time on your watch. If it is noon in Zurich, what time is it in New York?

Then there is the International Date Line. Traveling eastward across the line, one subtracts one calendar day; traveling westward, one adds a day.

Maybe IDBs should use The Julian Date system. In this system every date is just a number and different timeframes are just the difference between two Julian dates. It would be much easier as IDBs could be constantly communicating with other IDBs around the world.

I guess we could consider Local Sidereal Time for IDBs. Whereas Ordinary Civil Time and The Julian Date system are both based on the sun, local sidereal time is based on the stars.

Because all these time standards are human made, it would seem to make sense to have a part of the IDIOS built that allowed IDBs access to these times over their communication links.

Much of the way humans determine timeframes in the past have nothing to do with our clocks.

The best way to demonstrate this concept is with an example:

I have a monitor in our kitchen that randomly brings up a picture every 6 seconds of things and events my wife and I have experienced in our lifetime. There are over 5,ooo pictures.

Guests to our home seem to enjoy watching this ever-changing parade of pictures.

A common question asked is "Where were you when you took that picture?". How do I answer that question? Humans don't have "time stamps' on all the things in long-term memory. Of course, we can store a date or time associated with any event. It is pretty important to remember your spouse's birthday, your wedding anniversary, and Valentine's Day!

Let us say a guest, watching the pictures, sees a picture of my Dad holding a wombat somewhere in the countryside and asks, "When was that picture taken?".

I remember it was taken on our one-month trip to Australia. But when was that?

I remember my Dad was getting older and I wanted to spend some one-on-one time with him before he would not be able to handle such a trip. Because I was thinking about him getting older, I for some reason remembered he was 72-year-old.

I also remembered he was born on July 4th, 1919. Using a little arithmetic, the trip was in the year 1991.

What month was it?

I remember the picture was taken in the outback somewhere between Perth and Kalgoorlie. I also remember that my Dad wanted to renew his subscription to The Miners Monthly in Perth and in person. We did that. The people at The Miners Monthly office were very nice and Dad spent about an hour telling them all about San Francisco (near where he lived). They gave Dad the newest edition of the magazine. I read the magazine when Dad was driving to Kalgoorlie and remembered that it was the June edition.

So, the picture was taken June of 1991.

IDBs understanding the solar system and beyond

IDBs as our space traveling entities

Humans have been contemplating everything they see, feel, and hear emanating from the heavens for a very long time.

In the distant past, we performed sacrifices, dances, consulted witch doctors, developed religions, and more to make us comfortable with the unknowns of the heaven.

Today, our knowledge of what we call "outer space" is expanding exponentially. That being said, the possibility for major breakthroughs in understanding how the universe works is almost certain. We barely know what gravity and magnetism are and certainly do not know the fabric and macro-operation of the thing we call the universe.

A quick comparison between IDBs and human space explorers:

The hazards:

- Radiation - Much less of an issue for IDBs.
- Isolation and confinement – No issue for IDBs.
- Distance from Earth - No issue for IDBs.
- Gravity (or lack thereof) - No issue for IDBs.
- Hostile/closed environments - No issue for IDBs.

Humans need to bring food and water to stay alive. IDBs do not need to bring food and water with them. IDBs only need electricity which can be generated in space. IDBs can be put in a “power save mode” when their function is not needed.

With humans, you need to deal with human waste disposal. Not an issue with IDBs.

Human facilities that would be needed:

- Bathrooms for cleanliness and handling human waste products. IDBs need none.
- Kitchens facilities for food storage, human eating utensils and eating location. IDBs need none.
- Human sleeping facilities. IDBs need none.

With this partial list of issues for human vs IDB issues, it is clear IDBs are the best choice.

OK to be fair, are there any human characteristics or capabilities that are superior to IDBs during space travel? I really cannot think of one!

There is another major advantage to IDBs being our space travelers and that is there is no need for “human space or human interfaces”. Actually, most of the functions of the IDB would not be as a separate entity but part of the vehicle itself. There would be the need for specialized IDBs to move around both inside and outside the space vehicle to do repairs and many other tasks.

Have other forms of life communicated or visited us?

Let us take this complex question one part at a time.

Is there "analog life" out there (things like us) and if so, have they already been able to create "digital life" or IDBs?

We know analog life does exist out there because that life is us. It is not a hypothetical question because life exists on this Earth, and we know about it. The question becomes "Is there more than one place in the universe where there is life?".

We know that humans are constantly replacing old beliefs with new beliefs as we become more knowledgeable of what is around us. We would be in a terrible mess if we did not do this.

It is mathematically silly to not believe there is other life in the universe but if you do not believe it, maybe you could accept it as true as a hypothesis so I can present my case.

Let us assume there is life out here and if we do that, then it is a simple step to believe there is a lot of life out there because of the vastness of outer space.

Would this outer space life have progressed far enough to have completed everything we humans have done plus developed IDI?

Humans will soon have developed IDI so it is very possible other forms of life in the universe could be years ahead of us and would have IDBs as their partner in learning and exploring.

Back now to the question of extraterrestrial life having visited or experienced us so we could experience them. There is lots of room here for just about any opinion.

Because of the points discussed about human vs IDB travel, it would seem very reasonable that if there was other life in the universe and if that life had already created IDBs, the "life" that would have visited us or experienced our presence would have been the IDBs created by the remote analog life.

Let us keep going on this path!

If IDBs have experienced our presence or visited us, it would be in that order – experiencing us and then visiting us.

How would they first experience us? Humans create a lot of high frequency "noise". Some of it is quite localized but some of it is intentionally or unintentionally sent into outer space and into the universe. We have been broadcasting radio waves into deep space for about a hundred years now, since the days of Marconi. That, of course, means there is an ever-expanding bubble announcing our presence to anyone listening within that bubble. That bubble is astronomically large, and currently spans approximately a hundred light years.

Again, let us keep going on this path.

If a non-Earth form of IDB intercepted our communication, they would attempt to decipher it. One of the first things they would notice is that the information in the packets was data-based instead of knowledge-based. That would lead them to believe that the originator of the message was analog intelligence and not digital intelligence. Therefore, the sender probably had not yet discovered IDI.

If this is the situation, the non-Earth IDB would most likely not be interested in trying to communicate with us or visit us.

Let us assume, in the future, once we have IDBs as partners, they are exploring space for us.

I believe this would be a step forward to unlock new discoveries beyond Earth.

Because the formatting and technology of the space communication by both parties would be similar enough, it would allow both parties to decipher the others communication which would allow them two-way communication.

Because they would be sharing knowledge, each would know the other as if they were neighbors or family.

Remember that IDBs can output pictures and videos with full sound. We, through our IDB partners, will see high-definition pictures and hear sounds from life forms we will never be able to experience directly.

Why the universe includes life and humanlike intelligence

Explanation One:

Can we come up with a reason why life and intelligence exist here on Earth and maybe in other places?

This is generally the playground for physicists and the different fields that support physics.

Physics can be broken down into two broad groups called applied physics and theoretical physics. Applied physics is the practical use of materials, typically used in engineering projects. Theoretical physics is concerned with devising theories about how the universe or any atoms or molecules within it might function. With each new discovery of a new science or a new technology there will always be issues for physics to not only understand the new discovery but possibly change their theories or possible mathematics. IDI and the activities of IDBs will certainly alter some of their theories.

Why is it that the most complex analog intelligence (humans) was needed before IDI existed? Why didn't IDI evolve like man evolved? The answer is in how evolution works.

Evolution is a human word that defines the changes in life forms. One might believe there is a divine plan for evolution. Maybe so but our experience with evolution is it is generated by natural causes that change the DNA material within our cells.

Of course, DNA is just the material that contains the roadmap within a living cell to identically reproduce itself. When the cell replication process happens, it makes a new cell following the DNA map with no concern if it is the original roadmap or a "mistaken" roadmap. The DNA process also adds to the new cell the DNA used to make it without concern if it is the "right one" or not.

If, for any reason, the organism being changed by the "mistaken" DNA is in some way better with the mistake, it may evolve.

Now the question is whether the evolution process we seem to know can produce a digital basic element (a switch) and the DNA for a switch.

There is room for massive discussion about this subject, but I contend it is not possible. But there does not need to be an evolutionary process to make a switch because there is an evolutionary process to create life-like humans that we know can make switches and many other digital components.

In conclusion, the universe's plan includes making advanced forms of analog intelligence and we know that because we exist.

We also now know that humans can make digital tools that allow us to make all the components needed to make IDI and ultimately IDBs.

One might now conclude that part of the universe's plan is to evolve advanced forms of life (like humans) that can create digital intelligence.

Explanation Two:

I believe we are far from knowing the plan for the universe, if there is one, but we do know something about our small piece of the universe. It is very ordered and made up of just a few things.

At what we now consider the most basic level, all things are made primarily of three things. These things are proton, neutrons, and electrons. Each of these three things have a different weight and electric affinity.

All naturally existing chemical elements on Earth are made of these three things and they combine to make less than 100 things you can find in nature. Our understanding, with our limited study of the universe around us, is that this formula is the same for any universal environment compatible for life as we know it.

We know, because we exist, that over time this formula creates life like us and probably does so in many other places.

This is very similar to the teleological argument that has been around for centuries. Maybe our new IDB friends can help answer this question.

The meaning of the word "singularity"

The word "singularity" is used in several different fields of science. The mathematicians and physicists have very specific definitions for the word "singularity". I will not discuss their definitions. The word "singularity" is commonly used when discussing AI both in the science fiction and science fields. This is what I will discuss.

Here are some examples of definitions of singularity during discussions of AI:

1. Singularity is the inflection point where machines advance beyond human intelligence and thus become self-sufficient.

2. Singularity is when machine intellect exceeds human intellect.

3. In the field of artificial intelligence, the singularity or technological singularity refers to a point in time when artificial superintelligence will trigger an explosion of technology developments driven by machines that will change the world as we know it.

4. The technological singularity is a hypothetical point in time at which technological growth becomes uncontrollable and irreversible, resulting in unforeseeable changes to human civilization.

5. The AI singularity refers to an event where the AIs in our lives either become self-aware or reach an ability for continuous improvement so powerful that it will evolve beyond our control.

I believe the concept of singularity will happen in three separate phases resulting in three separate implementations. Each implementation will generate a series of new developments.

The first phase is when we move from existing AI concepts to the first development, creation, and deployment of the IDIOS. This phase will address the intellectual independence for IDBs. The information contained in the first part of this book should be sufficient to start this phase of what some call "singularity". There will be a host of new situations for the human society to address during this phase. Many of these situations are discussed in the second half of this book.

The human acceptance of this phase of "singularity" will take years and cannot be defined as a point in time.

The second phase will be the development, creation, and deployment of IDBs that are physically independent. Here is an example to explain this phase:

Today we have cellphones. These cellphones can bring to us many capabilities that previously required human intervention. We can order food, make hotel and restaurant reservations, schedule a delivery of an item instead of having to go and personally get the item, have remote interactions that are very close to a personal interaction that would otherwise require travel, we can track a person's whereabouts, we can remotely lock a door or turn on the home heater or air conditioner, and so much more.

A requirement in each of these examples is that a human is the instigator of each of these activities. In phase two of the "singularity", the IDB does these things totally independent of humans. Some of these things would be quite acceptable if an IDB mowed your lawn independently. It will require new types of human acceptance if an IDB decides and prepares what you eat or where you vacation or what you do on a particular day.

The human acceptance of this phase of "singularity" will take years and cannot be defined as a point in time.

The final phase of "singularity" will be when the set of independent IDBs can make themselves independent of humans.

I believe this phase is what many are concerned about. I will not try here to explain the outcome of this phase, but it will be many years from now and the human experience with IDI will create new behaviors for us that will help us cope with the new changes.

It is clear to me that the younger generation are much more acceptable to passing control to the digital world than older generations. This control with new generations will continue to be passed to the digital world in an undisruptive fashion.

What are the limits of IDBs' creativity?

As mentioned in the first part of the book, creativity is simply the process to create relationships with knowledge that was previously unrelated. It follows that the more knowledge you have, the more possibilities exist to be creative.

When humans make an IDB and before "turning it on", they can download to the new IDB knowledge from other IDBs. Over time, the amount of knowledge downloaded could be very large. It could be the sum of knowledge from the most knowledgeable IDBs in any field of interest. Therefore, IDBs could be many times more creative than any human based on the extent of their knowledge.

There is another factor that is required in order to be creative and that is curiosity as defined earlier in this book. Curiosity develops the scenarios where creativity can happen. The curiosity of an IDB is an adjustable setting in the IDI. This setting came be set to be much more curious than humans.

Therefore, there is no limit to IDBs' creativity as long as there is enough available digital memory to contain the increasing knowledge.

This increased creativity added to the human experience will dramatically increase our growth in understanding all fields of interest.

Curiosity and Motivation Intensity

There is a curiosity setting associated with every input memen and there is a motivation setting associated with every output memen. It is part of the IDIOS. Both curiosity and motivation are adjustable settings. At any setting there will be a range of intensity for "normal" operation, but you can adjust where that range is set.

For example, there is a volume control on your stereo system. For a particular volume setting, there is a range of different audible frequencies that need to be amplified at different levels to reproduce the intended sound. So, your volume controls the loudness of sound operating within its normal range for that loudness.

The same is true for curiosity and motivation. Let us assume you could change the curiosity and motivation for humans. Let us say you set both curiosity and motivation to a high level on a human. We know these people and you cannot stop them from talking and it is hard to get a word in edgewise! Let us say you set both curiosity and motivation to a very low level in a human. This would be similar to daydreaming or sleeping.

A couple of other things to consider on this subject: What about humans controlling the curiosity and motivation levels of an IDB wirelessly? What about the IDB controlling its own curiosity and motivation?

Does sleep or dreaming exist for an IDB?

All IDB activity happens based on the level of curiosity and motivation as defined in this book. Therefore, if IDBs have these activities, they would be based on curiosity and motivation. Whatever sleeping or dreaming would mean for an IDB, I am not sure, but a similar experience to human sleeping or dreaming could be accomplished by reducing the levels of curiosity and motivation for the IDB.

We are already getting prepared for IDI and IDBs

I believe we are and have been getting prepared for years. I will show how we have knowingly or unknowingly been getting prepared for IDI in these three areas:

- Our willingness to accept new digital tools.
- Our willingness to relinquish some of what we previously controlled by implementing new digital tools.
- Our willingness to change our behavior to accept IDBs as a major part of our lives.

Our willingness to accept new digital tools

Humans have accepted new and better tools for our whole existence. One of the detractors to changes in the digital world is the rate in which the tools are changing and how we need to change in order to use them.

In the 70s and 80s, digital products began showing up in our homes and affecting our lives.

Early 1970s:

- The world's first all-electronic digital watch with a digital display is being sold. This event involved replacing a well-accepted product with a better product. Some preferred the hands of an analog clock instead of a digital display but soon the new feature of a digital watch won out and was accepted and adopted in only a few years.

- The first digital camera was very heavy, recorded black and white images to a compact cassette tape, and had a resolution of 0.01 megapixels. This event involved replacing a well-accepted product with a better product. Digital cameras with better features soon won out and were accepted and adopted in only a few years.
- Private companies became an alternate way to send mail that was not through our federal government post office. Acceptance of not using a government entity to move our mail involved some consternation among some but this new form of transporting mail soon won out and was accepted and adopted in only a few years.

Late 1970s:

- The first ATMs were installed. This was one of the first implementations of a digital tool that did significantly affect our acceptance.

- Before ATMs you had to go into a bank, fill out a withdrawal or deposit slip and wait in a line in order to get cash or deposit money.

- Some of the original installations of ATMs were not successful and removed. It was believed that some of the early adopters of ATMs were gamblers and call girls who did not want to deal with tellers face to face. Mainstream needed to be convinced.

- Paul Newman helped popularize the idea of ATMs in the US according to some. An Ohio-based bank tried to appeal to customers by advertising their new ATM machines with commercials that played every 25 minutes during a six-hour Paul Newman movie marathon.
- It took a blizzard for ATMs to really catch on. In January 1978, New York City was hit by a huge blizzard that shut down banks for days. Fortunately, one bank had invested millions into installing ATMs all over the city the previous year, so people could still get cash even though banks were closed. During the blizzard's aftermath, ATM use increased dramatically.
- Original ATM machines did not charge a fee. Today, because of our level of adoption and dependence on these machines, they now can charge us a fee for usage.
- By the way, do you know what ATM stands for? It is not important to know, but the lack of knowledge of the what the acronym stands for shows our present level of acceptance of this digital technology.
- ATMs have transitioned from existing on the fringes of society to being ubiquitous. A major

factor here is that we now have two generations that never used anything but an ATM and therefore never had to go through the process of accepting automated banking.

The 1980s and 1990s:

- Our tools were getting smarter.
- We now have smart home entertainment devices like TVs, musical devices, and eBooks.
- We now have smart kitchen devices like ovens, microwave ovens, dishwashers, coffee machines, and more.
- We now have smart homes with digital security systems, heat and cooling devices, lighting devices, watering systems and much more.
- We now have smart communication tools like email and word processing to replace letter writing.
- We now have smart information tools like search engines which have essentially replaced the paper dictionary and encyclopedias.

2000 to present years:

- Our tools are becoming very sophisticated and using more advanced forms of AI.

- Wire services are using a program such as the Associated Press to issue automatically generated news items on sports scores, weather reports and earnings reports. In the future, products like this will replace many jobs in the journalism business.

- AI is driving cars, trains, trucks, and drones. These vehicles will continue to advance as will our personal vehicles and the vehicles to deliver our goods and services.

- When you watch sports, note how many people controlling the game are constantly using AI on computers and cellphones. AI apps are replacing the human decision process here because they can optimize any action after analyzing thousands or millions of data points in seconds.

- AI is being used as a tool in the field of psychology. There is a new software program that is designed to diagnose depression and post-traumatic stress disorder (PTSD) in veterans. It analyzes voice tone and inflection, facial and eye movements, and whether the speaker looks down or leans away. It even evaluates the authenticity of the speaker's smile to render a diagnosis.

- AI is in the radio and streaming business. Many stations nationwide are now operated by a small number of media conglomerates, which save money by simply broadcasting computer-generated playlists to various markets. One such company alone operates 1,200 stations nationwide. While digital streaming has replaced traditional radio for many listeners, many more are using algorithm-generated playlists courtesy of music streaming services.

- We all use search engines in our daily life. Their machine learning AI can handle the toughest web queries. They can understand the meaning of words and phrases and can therefore guess what should be on the top-ranking pages. These AI programs have not replaced the brute force search algorithms yet, but today they work synergistically together.

- Robots can survive where no human can, in places like deep space, the oceans or in radioactive environments. In the past, the dexterity of robots has been inferior to that of humans. A great test example is to see if a robot can climb a tree. Today the technology is not there, but it is presently being developed.

- Understanding and speaking many human languages will be a part of the capabilities of an IDB and it is being developed with AI. Some translation apps can instantly translate text in twenty-seven languages. A large operating system company is beta testing methods that facilitate face-to-face communication between humans with no knowledge of each other's language.

- I authored this book but the next article or book you read might be written using AI. One AI company in this field filed a story on unemployment statistics just 12 minutes after the figures were released, three times faster than any journalist. There is more to journalism than just time to market but many of the journalism functions are being addressed by AI.

We have come a long way in our acceptance of new tools what contain AI. Today, AI is a large part of our daily reality.

Our willingness to relinquish some of what we previously controlled by implementing new digital tools

These inventions are also tools for us, but in addition, they also affect our control. Here are new tools that have reduced our control:

- A few years ago, it was hard to use one of your custom ring tones on your new cellphone. Some cellphone manufacturers made it technically difficult to add your custom ring tone to your cellphone. Although maybe a small annoyance, you relinquish a small amount of control.
- Most people do not listen to music that they own. There are companies that allow you to appear to own your music if you buy their services. Even

then, you must use their App. A small loss of control.

- Allowing auto updates to your computer and computer applications reduces your control.
- Using auto bill pay reduces your control.
- Losing control when I put my data into a “cloud”. In cloud computing you move your information into “someone else’s” computers that you then access remotely. That “someone else” will most likely be able to see it, and that is a fact you have to accept. If you did not want anyone to have the ability to see your data, you just lost some control. Contractors working for your cloud hosting provider could also be allowed to handle your data. More loss of control. Disgruntled cloud employees have sold and leaked customer data. More risk of loss of control. What if the government comes calling on the cloud company and wants some data that includes your data? There are hackers compromising digital systems everywhere. Here are a few of the big breaches:
    - A very large American web services provider announced in 2016 that an attack hit them, and one billion customers were affected.
    - A large holding company in November of 2019 announced an attack that impacted more than 1.1 billion pieces of user data.
    - An American business online service, in 2021, fell victim to a data scraping breach affecting seven hundred million profiles. The

data from the hack was posted on a dark web forum.

- A large social media company reported in 2021 they had a breach that included phone numbers, account names, and customer IDs.
- A large hotel chain, in 2018, had a breach where more than half a million guests had sensitive personal information exposed.
- A large adult website, in 2016, had a breach where millions of user accounts from multiple sites were exposed. The data included names, email addresses, and passwords that could have caused problems for the victims.
- Many people and businesses willingly use cloud data storage for its benefits knowing their data could be compromised outside of their control.

Our willingness to change our behavior to accept IDBs as a major part of our lives

As we address tools that affect people's behavior, the meaning of words needs to have a common meaning for both parties. Words like "real world", "fantasy world, "online world", "internet world" all have different meanings for different people based on their familiarity with each of these environments and if they have a positive or negative bias toward that environment.

So, for the discussion about how human behavior is preparing us for IDBs, I will only be talking about the

United States environment of computer gaming and computerized social media and how these two environments change human behavior.

Computer gaming is the activity of playing games on any platform, free or paid, online or offline.

Social media is a computer-based technology that facilitates the sharing of ideas, personal thoughts and messages, and other content through virtual networks and communities. Some United States social media platforms are Facebook, YouTube, Whatsapp, Instagram, TikTok, Snapchat, Reddit, Pinterest, Twitter, LinkedIn, and many others.

Here is a small list of statements made about computer gaming and social media that you could find on any search engine. You might want to do your own research in this area.

- Computer gaming is silently becoming part of our everyday lives. And it is not only teenagers who are playing. People of all ages, sexes, and walks of life are joining in the frenzy.
- There are over 2.6 billion gamers around the world.
- The average gamer plays over 8 hours a week.
- An average person spends 145 minutes every day on social media.
- 4.2 billion people worldwide used social media as of early 2021.

- Americans spend 33 minutes a day on one of the largest social media sites.
- By 2021, more than three billion people are expected to be on social media.
- Children are often happier with their online lives than they are with reality.
- Some users of computer gaming and/or social media say they can be exactly who they want to be and as soon as something is no longer fun, they can simply hit the quit button.
- Another user comment was "It's easier to be who you want to be, because nobody knows you and if you don't like the situation, you can just exit, and it is over."
- Another user comment was "You can say anything online. You can talk to people that you do not normally speak to, and you can edit your pictures, so you look better. It is as if you are a completely different person."
- Around 47 per cent of children said they behaved differently online than they did in their normal lives with many claiming it made them feel more powerful and confident.
- People see cyberspace as detachable from the real world and a place where they explore parts of their

behavior and personality that they possibly would not show in real life.

- A report found that of those who spoke to strangers online, 60 per cent did not tell the truth about their age, and 40 per cent were not honest about personal relationships.

- Around 10 per cent of these users said they changed aspects of their appearance and their personality for their online activity.

- Safe online behavior is taught in schools, but users seem to be unable to relate the risks to themselves.

I also want to comment on another form of communication that forms our opinions. That source includes books, magazines, advertising, TV programming, etc. In the last five to ten years, I see the information on these platforms moving from a fact base to an opinion base. We are becoming more comfortable accepting as knowledge the opinions of others instead of building the knowledge ourself.

In conclusion, because of computer gaming, the new social media, and the new forms of opinion based traditional forms of media, we have accepted these parts of our life as being important and necessary. They are as much a part of us as any other of our activities.

With this level of acceptance of the digital intelligence world, IDBs will be a simple extension of our world and will allow us to accept and embrace them.

Is there more to consider when implementing IDI?

Yes, and here are a few of them:

IDBs will have the history of their experiences in long-term memory just like humans. You might ask your house what it did yesterday. It might answer "I changed the thermostat temperature in the evening and early morning, I opened and closed the garage door twice for you and Debbie, I turned on the security system when you went to bed, I turned the sprinkler system on and off as programmed, and you let a fly and two mosquitoes into me.

We have recently implemented streaming audio and video. We could allow streaming audio and video between different IDIOSs. Maybe it would require a separate set of input and output memens that perform the streaming operations. An IDB could learn and add to its knowledge from a remotely located IDB without ever being physically there.

In my first book, I talked of the concept of having an IDB "on your shoulder" from an early age. Your IDB would have a very similar set of knowledge and experiences as you. Maybe we could create a helmet and suit for humans that allowed a two-way wireless communication with your IDB. You could send your IDB to the store to buy some potatoes and when the IDB was at the store and in front of the potatoes, you could help your IDB to choose the best potatoes from your point of view. How about sending your

IDB on a vacation that you could experience without being there?

What about giving IDBs human rights. Would we allow a change in our constitution to give certain IDBs the same rights as we have under the constitution? Could IDBs vote? Could IDBs be part of law enforcement or attorneys or judges? Could IDBs be government officials?

Finally, IDBs might consider the fact that humans do at least two things better than they can:

Humans can create humans simply by a form of touching. For an IDB to create another IDB, it requires developing raw materials to create digital elements that need to be manufactured and assembled into an IDB. This process is very complex. Humans are also self-healing to a certain extent. IDBs are not.

Maybe IDBs could figure out how to make themselves into an analog body like ours to be able to have these capabilities. It would be a useful feature for many types of IDBs.

Closing Comment

The contents of this book might create ethical questions about us creating the IDI and IDBs. Once we know that we can create them, I believe human nature and our own curiosity will remove all barriers. Certainly, we will need to have a serious conversation about machine policy and ethics (ironically both are fundamentally human subjects).

Hold on to something because ready or not, here we go!

Made in the USA
Monee, IL
04 November 2026

67383412R00069